CREATION *or* EVOLUTION:
A LAYMAN'S LOOK

CREATION or EVOLUTION:
A LAYMAN'S LOOK

WILLIAM O' LEARY

PARTRIDGE

Copyright © 2021 by William O' Leary.

ISBN:	Softcover	978-1-5437-6503-8
	eBook	978-1-5437-6504-5

All rights reserved. No part of this book may be used or reproduced by any means, graphic, electronic, or mechanical, including photocopying, recording, taping or by any information storage retrieval system without the written permission of the author except in the case of brief quotations embodied in critical articles and reviews.

Because of the dynamic nature of the Internet, any web addresses or links contained in this book may have changed since publication and may no longer be valid. The views expressed in this work are solely those of the author and do not necessarily reflect the views of the publisher, and the publisher hereby disclaims any responsibility for them.

Scripture quotations marked NIV are taken from the Holy Bible, New International Version®. NIV®. Copyright © 1973, 1978, 1984 by International Bible Society. Used by permission of Zondervan. All rights reserved. [Biblica]

Print information available on the last page.

To order additional copies of this book, contact
Toll Free +65 3165 7531 (Singapore)
Toll Free +60 3 3099 4412 (Malaysia)
orders.singapore@partridgepublishing.com

www.partridgepublishing.com/singapore

CONTENTS

Daddy, Where Did We Come From? 1

One + One = ? 7

Let's Make a Snowman 11

Bombs Away 14

Bluster and Blowholes 15

Horsing Around 17

Monkeying Around 20

Methinks Me Smells a Rat! 22

Cake, Anyone? 24

The King's Clothes 27

Rhyme and Rhythm 29

Religion, Good or Bad? 34

Hiding in Plain Sight 39

Skeptic Thanks 44

Abandon Ship 53

Run, Rabbit, Run! 55

A Real Man 58

Small minds 61

Personal Testimony 65

Back in the 80s when I was very young, one day I was playing in an area that had a heap of rubbish nearby. Near the rubbish was a newspaper article that caught my eye, for some reason. It was about a group of scientists who thought they had witnessed the process of evolution happening right before their eyes. The article went on to say that while they were observing a rubbish heap, they witnessed flies appear from the rubbish. At first they were very excited, thinking they were witnessing the evolution of the fly. But on further investigation, they were disappointed to find that, instead of flies evolving before their eyes, they were only witnessing flies emerge from fly larvae that had been laid by adult flies.

That was the first time I had heard about evolution. But from the content of the article, I understood that these people were looking to prove that living creatures exist by appearing from out of nothing.

That was how I interpreted it at eleven or twelve years old. Four decades later, there seems to be no other way to interpret the theory of evolution in explaining how all matter came to be from nothing. I have a problem with that because it contradicts the laws of nature, math, and physics. Why would scientists latch onto a theory that contradicted all the laws of nature and could not be true unless there was a force outside of nature that could overrule the fixed laws of nature? On the one hand, scientists reject the idea of a creator because it can't be proven that such a being exists. On the other hand, they put aside everything they know, including plain common sense, and embrace a theory that has no grounding in provable fact. That's the same leap of faith taken by theists; but when people of faith declare a belief that God brought everything into being, including the laws of nature, they are mocked for not trusting in science as interpreted by those who reject the idea of a creator.

All things from nothing. Creationists and evolutionists subscribe to that belief. But while creationists believe God made all things from what can't be seen, evolutionists believe there is no creator behind creation but that all things are the result of a random big bang. The creationists (many of us anyway) believe God made all things in six days. The evolutionists believe all things came from nothing over a period of billions of years. I suppose their logic is, "Given enough time, anything is possible."

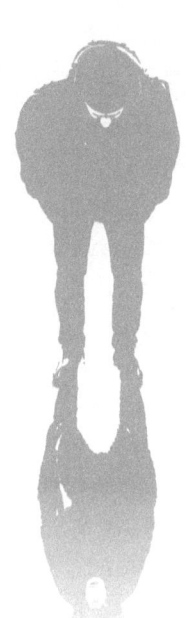

Daddy, Where Did We Come From?

One time a little girl asked her father, "Daddy, were did we come from?" Her father replied, "Well, God made us." The girl looked confused and said, "But Mummy said we came from monkeys." Her father thought about that for a while and said, "Oh, let me explain. Your mummy is telling you about her side of the family, and I'm telling you about my side of the family."

Maybe we should look at where the evolution theory came from in the first place. Without a doubt, the standout person behind the evolution theory would be Charles Darwin. The French had already worked out that species do change over time and become extinct. They just hadn't figured out how. But Darwin wanted to understand the "how." From 1831 to 1836, Darwin traveled to South America, Australia, and the southern tip of Africa. He was part of a survey expedition on the ship HMS *Beagle*. On his travels, he had access to hundreds of specimens from similar-looking species that lived close to each other but in slightly different environments. The most famous examples come from the Galapagos Islands off the coast of Ecuador. There he found many species, like iguanas, mockingbirds, and thrushes. But finches were the species he focused on the most and became obsessed with. Darwin discovered that finches had variations in how their beaks were shaped, which led him to believe that species evolved from more primitive forms.

On his return to London he published the *Zoology of the Voyage of* HMS *Beagle* in volumes from 1838 to 1843. These publications helped Darwin establish himself as a serious naturalist in the estimation of many. He not only became a member of the London scientific world but a leader in it. However, it was his five-hundred-page book *On the Origin of Species by Means of Natural Selection* that really made him a superstar. It became an instant best seller, planting the seeds for the theory of evolution that, two hundred years later, have grown to what it is today—the almost universal acceptance as fact that everything is here by chance.

Now we can look with more detail into what Darwin discovered and how he arrived at the conclusions he came to. In his book *On the Origin of Species*, he explains how descent with modification, or transformism, works. In any population of the same species, you can see a natural variation in traits—some finches have longer beaks; some have shorter beaks. Over time, small changes in the environment add up, favoring some traits over others. Natural selection modifies the population—the fittest survive and reproduce, passing on their traits, which over time lead to new species. He claimed that over many generations, differences in inheritable traits can accumulate in the groups to the extent that they are considered separate species. That, according to Darwin's theory of descent with modification, better known today as evolution, is how we all have come to be.

However, in order to have a new species, you need new information. If you cross a sheepdog with a German shepherd, you still get a dog, a variation within a kind from genetic diversity that was already there. There is no new information, no diverging into a new species. It has nothing to do with evolution; it's an adaptation. The Indian elephant has smaller ears than its African counterpart because it mostly lives in forests where smaller ears help it navigate its way through the dense trees more easily. The African elephant has larger ears because it lives on the plains and is more exposed to the sun. Their ears have thousands of tiny blood vessels. Flapping their ears helps them cool the blood in the vessels, sending cooler blood flowing throughout their bodies; otherwise, they could die from overheating. This is just one example of adaptation, not evolution, as Darwin claimed.

The finches in the Galapagos were finches before Darwin got there; they are still finches today. There is absolutely no proof that they evolved from other species or that they will evolve into other species in the future. In fact, Darwin admitted that himself when he thought about the seemingly fixity of species. He said that there was no evidence to show transitional forms. Of course, since then, evolutionists have come up with theories to explain these lacks of transitional forms; but there is no getting round the fact that no matter how they try to explain it away, finches are finches and elephants are elephants. Any explanation they put forward is theoretical and not based on observational or genetic facts at all.

Darwin's understanding was incomplete. What he observed was the different species adapting to their environment. For example: on one island, finches that ate large seeds tended to have large, tough beaks, while on other Islands, finches that ate insects tended to have thin, sharp beaks. The arctic fox sheds its white winter coat in summer; otherwise, it would stand out like a sore thumb, making it harder for it to hunt prey to survive.

Darwin had a problem with species seemingly being in a state of completion. It's only a problem when you can't accept the obvious and instead go looking for alternative answers. By virtue of your original premise being wrong, you then need to continually come up with increasingly complex theories that are not satisfactory at all and that no one really understands. When you refuse to accept obvious facts, you then end up having to do intellectual gymnastics to explain a theory that has no foundational truth at all.

In Corinthians 15, Paul, explaining the difference between the earthly body and the new glorified body, used as an example how "birds have one kind of flesh, animals another, and humans another." There we have it—common sense and simple observation explaining an irrefutable biological fact. All the so-called scientific statements and discoveries can't change the facts—there is no crossover of species and no proof that there ever was.

Let's first look at the evolution theory—random change, everything is here by chance! Some very obvious and logical questions come to mind when thinking about all things being here by chance. How did our world, our universe, and universes beyond just appear from nothing, without any plan or planner behind them? Then later on, much later on, even if some random chemicals mixed together by chance and a bolt of lightning did happen to strike those chemicals, giving birth to the first blob or single primitive cell from which all life supposedly evolved from, how then did that cell evolve, duplicate itself, and develop into what we are today—a life of incredible diversity and complexity? Could all of this happen by chance?

Most species, in order to continue to exist, need a male and a female. If, as evolutionists tell us, we are here by chance, then how did we evolve to where each species can reproduce after its own kind? Even in the case of nonconscious life forms like plants and trees, plants need birds, bees, and other creatures for pollination so they can reproduce. Trees produce seeds with the DNA to make exact copies of themselves. When an acorn takes root in the soil and grows to a giant oak tree that can live for up to five hundred years, it plays its part in aiding life by producing oxygen for other creatures and cleaning the atmosphere. Which evolved first, the acorn or the oak tree? There are thousands of different kinds of trees and other plant life. Did they all randomly evolve and then randomly evolve the ability to reproduce? Everything that has a beginning has a cause. If that's true of the universe, it is equally true of all life forms.

Evolutionists admit that there is design in nature. How can they not? But they reject completely the idea that there is a designer. To admit that there is design, but claim that there is no designer, doesn't make any sense. "Gucci products design themselves." I'm sure the owners of the Gucci brand would have something to say about that. In the same way, the Bible has something to say to those who claim there is no creator. Psalm 14 says, "The fool says in his heart there is no God."

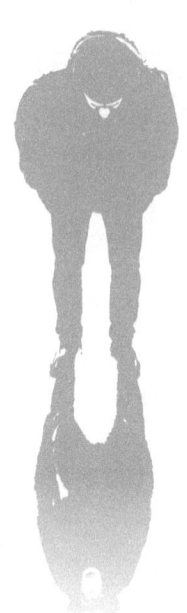

One + One = ?

Evolutionists say the earth formed about 4.5 billion years ago and started to cool down about 4 billion years ago. Some scientists believe that at one time there were about five million species on the earth. Dividing 4 billion by 5 million comes to 85. That means that there would be one brand-new species evolving every 85 years. On an evolutionary timescale where thousands of years is like the blink of an eye, that's warp speed. Evolutionists would say that that kind of phenomenon happened only twice, one million years ago during the Cambrian Era and again when man first came down out of the trees, about two hundred thousand years ago.

The evolution of the car, you could say, began with the invention of the wheel. The first motor vehicle built by Mercedes-Benz in 1904 had a lever for steering, a flat floorboard, and less power than the average grass mowers of today. The modern car of today, by comparison, is made up of about twenty-four thousand parts and is light years more advanced than that first car. The Car could never have evolved by chance; and yet people choose to believe that creation, especially conscious life forms that are almost infinitely more complex than any mechanical machine, are here by chance.

The idea is that all life, including mankind, evolved from a single primitive cell. However, biology has shown us that of the 10 billion or so cells that make up the human body, every one of them contains the DNA to make an exact copy. Science has shown us that life is far more complex than we could have imagined even only a hundred years ago. There's not a single primitive cell in the human body, and it seems there never was.

The human body is perfectly adapted to live in earth's environment. We have skin to protect us from harmful viruses and bacteria. We have lungs that help us to live, giving oxygen that's transported around our bodies by the pumping action of the heart through about two hundred miles of veins, keeping us healthy and alive. We have eyes to see and enjoy the beautiful world around us and a brain to process and understand all that we see. We have saliva in our mouths that help us to start breaking down food even before we chew and swallow and send it through the processing plant that is our stomach. Then it passes out of the body through a very efficient waste disposal system. We have kidneys that filter out harmful poisons that would quickly kill us if our kidneys were not as efficient as they are at doing the job they were created for. These organs are so incredibly complex, far more complex than any machine or device ever created by man.

Take the eye, for example. Light enters through the cornea (front glassy part of the eye), then through the lens, and onto the photoreceptors of the retina (back of the eye). Impulses of light converge on the optic nerve and then go to the brain. No computer chip today could begin to do what your retina does. If such a computer chip were made, it would have to be something like half a million times bigger than your retina. It would weigh about 100 pounds. The retina weighs less than a gram. It occupies only 0.0003 inch of space and uses only 0.0001 watts of power. A scientist's theoretical "seeing chip" would fill 10,000 cubic inches. It would require 300 watts of power and a cooling system. Even with all that, the retina can see five times more units of vision than a seeing chip. Such a chip would have the equivalent of about 1 million transistors, while the retina has the equivalent of 25 billion. Einstein said, "The eye is too complex to have evolved." Psalm 139 says, "I am wonderfully and fearfully made." It's only in the last few centuries that we have been finding out just how wonderfully and fearfully we are made.

Far from proving that a divine creator does not exist, science is showing us just how more and more unlikely it is that everything is here by chance and more likely that there is an intelligent designer behind creation.

Creation is far too complex to have just evolved from nothing, with no plan or designer behind it. It would be the same as someone looking at the Empire State Building and claiming it gradually evolved to what it is today. All the concrete, metal, wiring, and plumbing, not to mention the foundation, all randomly came together? You would be considered a first-class idiot to make such a claim. And yet, that's what the best and the brightest tell us is how we are here today— by chance. What does the Bible say about creation? "God made them male and female. Both male and female He made them." Straightaway, that makes more sense than all that has gone before. Genesis says the earth was covered with water. God separated the water, creating dry land. Then he said, "Let the earth sprout vegetation, plants yielding seed, and fruit trees bearing fruit *after their own kind*, with seed in them." Then God created the light so that the vegetation could live and thrive. Without light, all things would die within two weeks. God made the light in one day. First, he made the light and the vegetation to create the environment for life. Next, God created the fish of the sea, the birds of the air, and the land creatures. A very important verse says, "He made them, 'after their own kind.'" Even dumb animals know to which category or group they belong. There are no crossovers of species. Finally, God made man on the sixth day.

Let's Make a Snowman

The creation story gives a commonsense foundation for all life. The foundation of any idea or structure must be sound, but if not, what you build won't hold up. That all life came from a chance happening billions of years ago does not make any sense at all, if you really give it some thought. Take that story to any Hollywood producer and they would probably say, "Come back when the story is more plausible and believable." And yet, we are being told this is how all life began and why we are here today.

There is no way so much diversity in creation just evolved randomly. When you look at the order in nature and the sequencing of the genetic code and still choose to believe that everything—an almost infinite number of random events—is here by chance, that's stretching logic beyond reason. It would be like taking a bag of marbles, breaking the bag, and letting the marbles fall to the ground. They would roll off randomly in every direction.

If you did that a thousand times, trying to have the marbles fall in such a way that they would line up and form the word *evolution*, that would never happen, not even if you tried it a million times. Creation is, I will repeat, near infinitely more complex than that.

If we believe everything exists by chance, we cannot, in good conscience, change the rules and claim that somewhere along the line, random things began to organize themselves. We can fool ourselves and want it both ways when it suits us; but the reality is, it's either chance or something else, something planned. Make a snowball, and push it randomly down a hill. Following it down the hill, you will find that the snowball is much bigger than at first. But if you followed it down the hill and found a snowman, complete with hat, scarf, button eyes, mouth, carrot nose, and stick hands, then the only conclusion you could logically come to is that it was made by someone. If you follow logical thinking, there is a plan behind creation. It doesn't take a scientist to figure that out. It is common sense.

Our amazing blue earth is diverse in plant life and in conscious life, both animal and human life. We can see many examples of diversity—the periodic table of the elements, centrifugal force (the force that causes the planets to rotate in orbits and not spin off out of control), gravity (the force that causes the moon to influence the tides here on earth), etc. How can anyone claim that all this is by chance? You would be more likely to win the lotto a hundred times in a row than to have all this happen by chance! Such thinking defies logic and common sense. And no, not all things are possible if given enough time.

There must be planning and materials in order to build something. The famous physicist Stephen Hawking once said, "The elements that make all life possible and allow for the existence of all things, are so finally balanced that if you could put a ruler across our universe and change them just an inch on the scale, nothing could exist." To paraphrase him, when everything came into existence, it had to be complete—all the elements perfectly in place. Otherwise, there could be no you, no me, and no world.

Bombs Away

Whenever there is an explosion of some kind, gases are created by the explosion. The highest concentration of gases occurs near where the initial explosion happens. As the explosion expands outward, the concentration of gases becomes less. Scientists who studied the beginning of our universe expected to find the highest concentration of gases there. Instead, they were amazed to find that the concentration of gases throughout our universe is constant.

If the universe exploded into being, it did so with incredible precision—it would have been a controlled explosion in an absolute sense. Stephen Hawking once said: "The universe would not exist if there was a decrease in the expansion rate one second after the Big Bang by one hundred thousand million, millions." He concluded, "It would be very difficult to explain how the universe came into being this way except if it was by a God who intended to create beings like us."

Bluster and Blowholes

A man went into a bar. He saw sitting at the counter two very large women speaking with an accent. He said, "Are you two lovely ladies from Scotland?" One of them said, "Wales, you idiot. Wales." He said, "Oh, excuse me. Are you two whales from Scotland?"

The proponents of the evolution theory go to great lengths to exclude the creation story. For example, evolutionists claim whales are descended from cows. But think about that for a while, and ask the question, "How much of a transformation would a cow have to undergo before it could change into a whale?" The famous mathematician Dr. David Berlinski did just that. Think about it. In fact, he went further and applied math to test the theory. He did the calculations and stopped at fifty thousand. He said, "Virtually every feature of a cow would have to be changed. Things like the breathing apparatus, the visual system, the skin would need to become waterproof, and the stomach would need to change completely. Cows eat grass, whales don't. And all the changes would have to happen simultaneously." It would be like a cow falling off a cliff, going down into the water, and coming up a whale. It's impossible! Yet that's the theory to explain the existence of whales. It seems it's not only mammals like whales and dolphins that have blowholes.

None of this is provable; it's not science. Neither is it rational. If you follow that kind of reasoning, why not suggest that we fling a human into space from a giant catapult, say, once a day for the next ten thousand years or so in the hopes that eventually man will evolve so that he can live in space. I suggest that we start with evolutionists first, then politicians, and after them, lawyers. At least, that way, some good would come from our madness, unlike the evolution theory, which has not benefited mankind in any way at all.

Horsing Around

However, evolutionists claim that there is fossil proof of some creatures evolving through different stages to the present. One frequently quoted example of this is the evolution of the horse:

Eohippus, the first horse, also known as the dawn horse—ten to twelve inches high, three toes in the back, and four toes in the front—lived 50 to 60 million years ago.

Miohippus—middle horse—size of a sheep, bigger teeth, 40 to 50 million years ago.

Mesohippus—slightly bigger, outer toes much smaller, long slender trunk, eyes further back, 26 to 40 million years ago.

Parahippus—side toes bore little weight, head and teeth much larger, 23 million years ago.

Merichippus—forty inches high, body proportions identical to today's horse, outer toes almost disappeared, center toe almost like a hoof, 17 million years ago.

Pliohippus—only one toe, like horse today, 15 million years ago.

Equus—fifty-two inches high (thirteen hands), large head, heavy neck, hair stood up like a shaving brush, the true horse today, ten thousand years ago to today.

Evolutionists also claim that the various fossils were found at different layers in the rocks with Eohippos being found furthest down, in keeping with its status as the first ancestor of the horse.

Here evolutionists make a lot of assumptions. As noted above, Eohippus—the so-called first horse, or dawn horse—was ten to twenty inches high, had three toes in the back, and four toes in the front, or the other way around. I don't know, but what I do know is that it doesn't sound like a horse to me.

Next up, we have Miohippus, who had a long, slender trunk and whose eyes were further back. It sounds more like an anteater or a miniature elephant.

Then Parahippus lost the trunk and managed to grow a bigger head and teeth. I wouldn't like to meet that thing on a dark night. I'm going to make a suggestion. What we have here is not the evolution of the horse at all but rather other distinct and separate species that are now extinct.

Here's what I found. Horses come in many different sizes, perhaps a greater range of sizes than any other land animal in the world. The largest horse is the Shire, which stands at seventy-two inches tall or about six feet. The smallest horse in the world is the Falabella, averaging about twenty-five to thirty-four inches. Their ancestors were like them, horses—not anteaters, miniature elephants, or something else.

In her book *Bone Of Contention*, Sylvia Baker states, "It is unlikely that Eohippus is an ancestor of the horse for the following reasons: A. Horse fossils are not found below one another in the rocks. On the contrary, bones of Eohippus are often to be found at the surface. B. The fossils are represented as showing a gradual increase in size as Eohippus developed into the modern horse. However, this type of argument is invalid since horses of many different sizes exist today. C. Eohippus and the modern horse both have eighteen pairs of ribs. But of the supposed intermediate forms, Orohippus had fifteen; and Pliohippus had nineteen. D. It is not generally realized that Eohippus has a skeleton that is very similar to that of the present-day creature known as the hyrax. Some scientists believe that Eohippus has no connection at all with the horse but is simply a variant form of the hyrax."

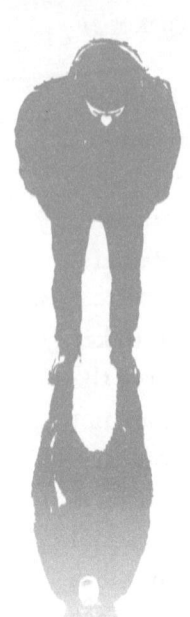

Monkeying Around

Archaeologists can tell us with certainty where civilization began—in the Middle East. Dig a hole just about anywhere in the Middle East, and you will find traces of the earliest origins of man. But show me the tree where man first came down out of. In the last hundred years or so, archaeologists have found that ancient Egyptian and Babylonian cultures were far more advanced than at first thought. They had large cities with multistoried houses, sewage systems, and underground heating in their homes. They had universities and were advanced in the sciences like geometry and trigonometry. That's what the archaeologists have found. The Bible gives us detailed records and names of these civilizations and, just as importantly, accurate times and dates. Abraham lived around 4000 BC when Babylon was in the process of becoming a rising power. Moses lived about 3000 BC when Egypt was a superpower. So archaeologists and evolutionists had a problem. If they were to accept the Biblical time frame, they would have to accept that these advanced civilizations sprang up suddenly, seemingly in a very brief period of time, out of nowhere. This does not fit their narrative that the earth is billions of years old and that man evolved from monkeys many thousands of years ago. How could they get around this? They completely ignored the information given in the Bible about the time frame and origins of things. Instead, they had man come down out of the trees about two hundred thousand years ago and begin to organize and build cities about one hundred thousand years later. This contradicts the biblical claim that states that the city of Babylon dates back about 4000 BC.

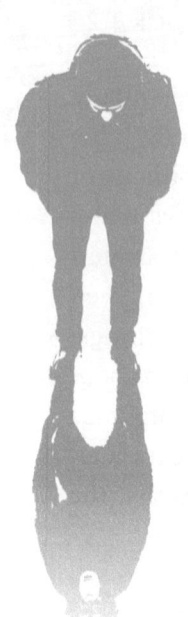

Methinks Me Smells a Rat!

It says in Genesis "God brought all the animals to Adam to see what he would name them." God ranked man first in the order of creation, gave him superior intelligence, and made him ruler over all of creation. More importantly, God gave him the spiritual faculty to know and love him.

In contrast, evolutionists rank man as just a subspecies who, for some unexplained reason, left all the other species in the dust, came down out of the trees, and advanced to the point where we can fly to the moon. All of this happened in only one hundred thousand years, a blink of an eye in evolutionary terms. Richard Dawkins explained it by saying, "Monkeys typing on typewriters, sooner or later would type the phrase, 'Methinks it is a weasel,' a quote from Shakespeare." Commenting on that, the famous mathematician David Berlinski said, "'Methinks it is a weasel' is a six-word sentence containing 28 English letters including the spaces. It occupies an isolated point in a space of 10,000 million, million, million, million, million, million possibilities" (Translation: "Richard, methinks it's you are that weasel").

Mr. Berlinski goes on to say, "Any definition of natural selection must plainly meet what I have called a rule against deferred success." It's not as if Mr. Berlinksi is ideologically biased against the evolutionist claims. He is coy on whether or not he believes an intelligent being is behind creation. I suspect he would be offended at the thought of there being anyone smarter than he, and for that reason, he would be hindered in believing.

From listening to his lectures, he seems to be a man who is guided solely by reason and the laws of nature and physics. I think that's a flawed approach because, at one time, none of these things existed. Creation, lifeforms, gravity, the elements that make up the periodic table, and all life and matter at one time didn't exist. The temporary can never explain the eternal. But he doesn't come across as hostile toward the idea of a creator. And that, I think, helps him to be more open and freer in his thinking when it comes to assessing how all things came into being. He is a man who looks to science and reason to explain creation; and he has, after applying his considerable powers of reason and mathematical ability, come to the conclusion that the evolution theory comes up way short in explaining life and matter.

Cake, Anyone?

We are told that coal is formed in swamps, with each layer taking thousands of years to form. Think of a layered cake, like a sponge cake. First a sponge (coal seam) layered with jam (sediment), another sponge (coal seam) layered with cream (sediment), and so on. Oftentimes, tree trunks have been found growing through a coal seam, through the overlying sediment layers, sometimes even into another coal seam. In other words, it is like a knife going through the sponge, cream, jam, on into the other sponge, and so on. If each layer took many thousands or millions of years to build up, it would be impossible for a tree to live that long. A tree would rot away in only a fraction of the time it took for a small layer to accumulate. Far from taking millions of years to develop, layers found with logs going through them are consistent with sediments having formed quickly by the action of lots of water.

As for coal taking thousands or even millions of years to form, actually coal can be formed in just weeks. All it takes is the right amount of pressure and naturally occurring ingredients, and coal can be made in a laboratory. The same is true with diamonds. Using certain materials and applying great pressure, diamonds can also be made in only a few weeks. The quality of the lab diamonds are so good that it takes an expert to tell the difference.

Fossils form very quickly, much more quickly than evolutionists would have us believe. In New Zealand, sausages, a bowler hat, and a leg of ham, which were buried in the 1886 volcanic eruption of Mount Tarawera, were all turned into stone.

We haven't even looked at the science of cells yet. Biology is much more complex than Darwin could have possibly known in his time. Professor of biology Michael Denton, in his book *Evolution: A Theory in Crisis*, states, "The cell is the most complex and most elegantly-designed system man has ever witnessed.

To grasp the reality of life as it has been revealed by molecular biology, we must magnify a cell a thousand million times until it's twenty kilometers in diameter and resembles a giant airship large enough to cover a great city like London or New York. What we would see would be an object of unparalleled complexity and adaptive design. On the surface of the cell, we would see millions of openings, like portholes of a vast spaceship, opening and closing to allow a continual stream of materials to flow in and out. If we were to enter one of those openings we would find ourselves in a world of supreme technology and bewildering complexity ... (a complexity) beyond our own creative capabilities. A reality which is the very antithesis of chance (in other words, no chance) which excels in every sense anything produced by the intelligence of man."

That a structure like the cell, of which our bodies have about 10 billion of them, could have just "evolved" does not make any sense. It only makes sense when you factor in a designer. David Meyers says, "Evolutionists put forward the theory of 'bottom up' materialist accounts where the molecules get more and more complex and form more complex molecules and cells, and the cells compete to form more complex organisms. And yet, what we see in life, complex miniature machines, information processing systems, digital code, these are things that bear the hallmark of mind. And they suggest rather a top-down instead of bottom-up approach."

The dictionary definition of the word *theory* is "a proposed explanation whose status is still conjecture." In other words, "your best guess in the absence of definitive proof." The evolution theory is just that, a theory and not a fact, not real science, which can be proven experimentally. In fact, when it comes to scientific progress, evolution has often been a hindrance because, unlike actual science, which can be proven in a lab, making assumptions based on theories leads to errors.

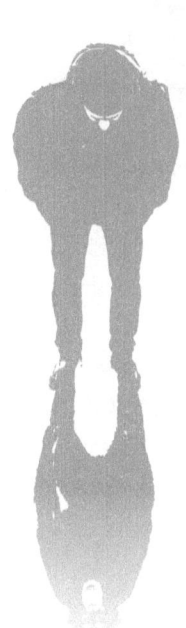

The King's Clothes

The biblical account of creation is soundly rejected by most people in academia. Evolution is the accepted dogma, amounting to a doctrine. There is absolutely no room for the biblical account of creation.

But in relation to how all things got here, evolutionists offer a less than satisfactory explanation on just about everything. Another example of that would be their theory as to how we have water. They claim the earth exists because 4.5 billion years ago meteors crashed together, forming the earth. For the first 500 million years, the earth was just a big ball of fire. But after coming up with that theory, they then had to explain the existence of water. How do they explain the presence of water on a fiery earth? Well, for about 2 million years, the earth was bombarded by asteroids and comets that carried tiny drops of water. That's the evolutionists' explanation. Compare their explanation to the Bible explanation in 2 Peter 3:5: "They deliberately ignore the fact that by the Word of God heaven existed long ago, and an earth was formed out of water and by means of water."

Considering that 70 percent of the earth is covered by water and the Mariana Trench, the deepest part of the seas, is about seven miles deep, the biblical explanation as to how we have water on the earth seems to make more sense.

An article in the *Irish Times* dated Thursday, January 5, 2017, written by a professor in the UCC named William Reville, tells about a discovery by Steve Jacobson of Northwestern University in the USA. Mr. Jacobson and his team discovered a huge reservoir of water 700 km to 1,000 km deep in the earth's mantle. The reservoir is so vast it contains more water than can be found on the surface of the earth. In Jacobson's words, "We should be grateful for this reservoir. If it wasn't there, it would be on the surface of the earth; and the mountain tops would be only land poking out."

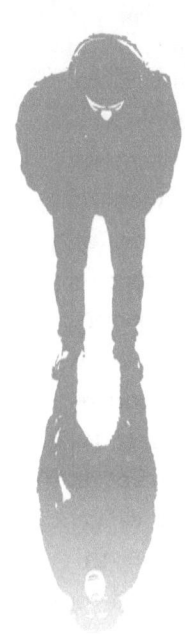

Rhyme and Rhythm

Nature is a living thing, but it doesn't think or feel. There is rhyme and rhythm to nature—the sun rises and sets; seasons come and go. Because of this consistency in nature, we can observe and record occasions like births, deaths, and everything in between. We have access to recorded history going back hundreds, even thousands, of years and with great accuracy. We can do this because nature is predictable.

Nature is very beautiful. Think of snow-capped mountains, valleys, and hills or the earth as viewed from space. Nature is alive in the sense that there are living organisms, like plants and trees, and conscious organisms, like animals and even higher life forms such as humans. So it's no wonder when many people speak about nature, they refer to it as Mother Nature. Some cultures worship nature. The Indian tribes of North America worship the four forces of nature earth—wind, water, and fire. The aboriginal tribes of Australia have what they call the dream time when they worship

Mother Nature, the giver and sustainer of all life. Even in this, the twenty-first century, we still have people who worship Mother Earth. Paul McCartney, the famous singer of the Beatles, has a song in which he refers to himself as Mother Nature's son. At certain times of the year in many parts of the world, people gather to worship and thank Mother Earth for seasonal harvests and to pray for future harvests. It's partly a way to try and control the future and partly to appease nature because nature can be so unpredictable and dangerous.

It's true we are part of nature. About 70 percent of our bodies are made up of water. When we die, we go back into the ground and become one with the earth. Everything we need to survive—oxygen, food and so on—we get from creation. When properly managed, nature thrives, allowing us to thrive with it. When we abuse nature, we suffer. Think of poor soil management, which leads to drought, or of overfishing, which leads to food shortages and loss of livelihood.

Nature is so prolific. Experts tell us that if we stopped fishing the oceans for one year, stocks would recover so spectacularly that we could fish them for a hundred years and barely make a dent into the fish populations.

But nature cannot decide whether to do us good or bad. It's not a moral entity that can judge good or bad, right or wrong. It's not nurturing in the way that a mother nurtures and cares for her offspring. Nature cannot endow us with something it doesn't have. We have a conscience. Nature is a force. It doesn't have feelings, hopes, or dreams. We have the ability to know right and wrong; we have hopes and dreams. We can make moral judgments. That ability comes from somewhere other than nature.

Genesis says, "God made man from the dust of the earth." We are made in God's image and likeness, which means we can create, like our Creator. Think of the Sistine Chapel in Rome, the Mona Lisa in France. God, the maker of creation, has endowed his creatures with creative abilities. Although our moral ability has been warped by sin, we still know right from wrong. This ability we get from God, the great moral law-giver. Is Mother Nature our great benefactor? Does Mother Nature know the future and know our destiny? How can that be? How can an impersonal force have such powers? Logic and reason say, "No. That cannot be so." To think so and to look to nature for that kind of help and assurance is foolish and fatally flawed.

There is wisdom and knowledge in nature that has clearly not been learned or acquired by nature itself. As was already said, it certainly does not know or judge right from wrong. Nature is a living, breathing thing; but it cannot do like God did when he looked at all he had made, and as Scripture says, "He saw that it was very good." Nature is like a cat that licks the hand of its master. It doesn't think great thoughts. It doesn't ask, "Why am I here? Where did I come from?" Like cats, it just does what it does because that's what it was created to do. In the same way, nature does what it was created to do. It does not ask or question why. It doesn't have the free will to question. It just is what nature is.

On the other hand, God not only has a plan for you and me, but he has a plan for nature as well. First, let's look at what Paul says about nature in its present form. Romans 8:19 to 23 states, "The creation waits in eager expectation for the sons of God to be revealed. For the creation was subjected to frustration, not by its own choice, but by the will of the One who subjected it, in hope that the creation itself would be liberated from its bondage to decay and brought into the glorious freedom of the children of God. We know that the whole creation has been groaning as in the pains of childbirth right up to the present time. Not only so, but we ourselves, who have the firstfruits of the Spirit, groan inwardly as we wait eagerly for our adoption as sons, the redemption of our bodies."

Second Peter 3:10–13 has this to say about the future of the earth: "But the day of the Lord will come like a thief. The heavens will disappear with a roar, the elements will be destroyed by fire, and the earth and everything in it will be burned up. Since everything will be destroyed in this way, what kind of people ought you to be? You ought to live holy and godly lives as you look forward to the day of God and speed its coming. That day will bring about the destruction of the elements by fire, and the elements will melt in the heat. But in keeping with His promise we are looking forward to new heavens and a new earth, the home of righteousness."

Think of atom bombs. The whole universe is made up of atoms. There is enough fire power to destroy our earth, our whole universe. Scientists can't argue with that. They estimate that, some day, our sun will implode and then explode, destroying our universe in the process. They say that's about 2 billion years away. Whether or not that's true, we don't know; but we do know what Scripture has to say about the destiny of our planet, and more importantly, our destinies. We sell ourselves short when we foolishly and blindly look to nature for security and survival. Nature can offer us neither, and we insult our Creator when we ignore him and instead look to his creation to give us what only the Creator can give. There's a price to pay for that—a gnawing emptiness in our hearts and a desperation to cling to this life because we have no assurances for the future. Without God, all we see before us is blackness. To quote the lyrics from a John Michael Talbot's song:

> Only in God will my soul find rest, in
> Him is my salvation.
> My stronghold, my Saviour.
> I will not be afraid at all.
> My stronghold, my Saviour.
> I will not be moved.

This kind of assurance you can never get from nature or from the things of time. Only Christ and give that assurance.

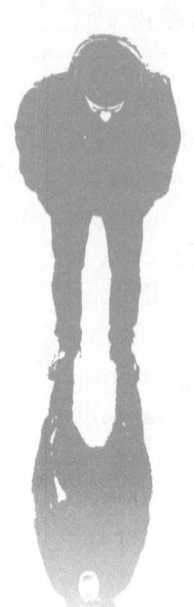

Religion, Good or Bad?

I have an atheist friend whom I have been having discussions with for close to thirty years now. He is, for the most part, a fair-minded and reasonable person until, that is, it comes to religion. Then all open-mindedness and fairness go out the window. Instead, he becomes dogmatic, angry, and almost religiously self-righteous.

One time we were discussing the Genesis Flood story. His response was to suggest that God has anger-management problems. I pointed out that there were about a hundred years between the beginning and the end of the construction of the ark and that by the time the Flood came, only eight people had heeded God's warning and entered the ark. Far from having anger-management problems, God was very patient and gave people many chances. In fact, one hundred years of chances. His response was silence for days, and then he moved on to other so-called inconsistencies in Scripture.

A man can love football. He can go to all the games of his favorite team, follow them on social media, and have his room covered in posters. Most people would say, "Ah, he's harmless. He just loves his football." A man can be religious, devoting his life to God, and many would say, "He is extremist and unbalanced."

Others cite the hypocrisy they often see among religious people as a reason that they reject all religions. Yes, a lot of bad things have been done in the name of religion, but that doesn't mean all religion is bad. It just means that bad people take something good and use it for their own selfish ends and, in the process, discourage people who otherwise might be open to experiencing the rich blessings that religion can bring.

James 1:26 to 27 states, "If anyone considers himself religious and yet does not keep a tight rein on his tongue, he deceives himself and his religion is worthless. Religion that God our Father accepts as pure and faultless is this: to look after orphans and widows in their distress and to keep oneself from being polluted by the world." Religion must be of the heart, not just outward and superficial.

In Acts, we read about a Roman centurion named Cornelius who was a good man and very religious. He loved God, helped the poor, and tried to live a good moral life. One day while he was praying, an angel appeared to him and told him to send to Joppa for a man named Peter, who would show him how to be saved. Four days later, Peter arrived and told an expectant Cornelius and his family and close friends about Jesus, about how he went about doing good, and how he died on the cross to pay for our sins. Cornelius and all his friends and family accepted Jesus and were saved. (The word *saved* in the Bible means to be in a right relationship with God when we put our trust in Jesus and his finished work on the cross.) Religion without a knowledge of the truth can be very dangerous and harmful. Paul was a member of the Pharisees, a very religious sect. But that wasn't enough for Paul. He says of himself that he surpassed many of his contemporaries when it came to obedience to his religion. In fact, he was so zealous that he went around arresting and torturing the followers of Jesus. He was filled with hatred for anyone who professed faith in Jesus. However, on the way to Damascus, Paul had an encounter with Jesus that changed his life. God softened and changed his heart so that, from then on, he loved all people and sought to tell them the message of new life in Christ. For the rest of his life, Paul was hounded and persecuted by the Pharisees, the religious group of which he was once a member. They were the same religious people who persecuted Jesus and eventually had Jesus killed. Yes, the most religious people of the day were the ones who had Jesus killed. So you see, religion without knowledge can be blind and very dangerous. I wouldn't like to be those people on Judgment Day.

I repeat, religion without knowledge is dangerous. Good intentions are not enough. A wise person once said, "The road to hell is paved with good intentions." How can you tell when a religion is good or bad? That's a little complicated because a religion needs to be based on truth to be really good. You can have good deeds and be sincere but still be in error.

There are many religions in the world. But in the interest of time and to keep things simple, let's look at the three main religions: Christianity, Buddhism, and Islam. To find out what a religion is all about, you should examine what the leaders say and how they lived. Buddha never claimed to be a religious leader; he was a philosopher. But later on, his followers added a spiritual dimension to his teachings. I suppose this was to satisfy their need to worship a higher power. Mohammad claimed to be a prophet of God and nothing more. But Jesus claims to be the eternal Son of God, who existed before anything came into existence. He claims to be the Way, the Truth and the Life. Anyone who believes in him has eternal life. Now, either he is lying, crazy, or telling the truth. It's unlikely he would lie; he is not crazy. So he must be telling the truth.

At this point you might argue, "Well, I can be a good person without having to identify with any religion or declare faith in God." Remember we talked earlier about Cornelius the centurion? The Bible says he was a good man who feared God and helped the poor. Yet the angel told him, "Send to Joppa for Peter, who will show you how to be saved." Peter brought to him the saving knowledge of salvation through faith in Jesus, God's sinless Son, who died to pay for our sins and make us right with God. Therein is true religion— it's faith and trust in Jesus, who died to pay for our sins and rose again to give us the hope of eternal life. Real religion is about relationship, a relationship with God through his Son, Jesus.

Many people claim to be followers of Jesus. However, they are nominal Christians and not true followers, who say no to carnal, worldly ways and who don't live by the Spirit, as Scripture tells us to. That hides the light and turns people off the Gospel. Gandhi once said, "Your Jesus I like. It's you Christians I have a problem with."

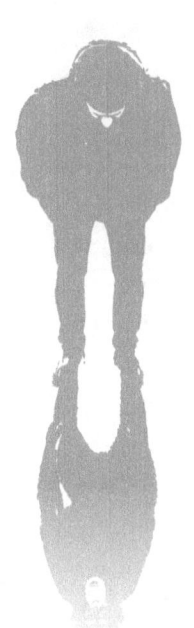

Hiding in Plain Sight

In the biblical account of the Flood, Scripture records in Genesis 7:10–12, "And it came about that after seven days, that the water of the Flood came upon the earth. In the six-hundredth year of Noah's life, in the second month, on the seventeenth day of the month, on the same day all the fountains of the great deep burst open, and the floodgates of the sky were opened. And rain fell on the earth for forty days and forty nights."

The flood story is not unique to just Jewish culture. Scholars estimate there are over five hundred flood legends worldwide. Ancient civilizations such as those of Scandinavia, China, Babylonia, Ireland, Wales, America, Russia, and so on, all have their own version of a great flood. The flood tales of these ancient cultures have many details in common with the biblical account, including the warning of the coming flood, the building of a boat, the storage of animals and people, and the sending out of birds to test if the water level had subsided. The overwhelming consistency among flood legends found in distant parts of the world indicates they were derived from the same origin, but oral transcription has changed the details over time.

Many claim that because the flood story is not unique to Jewish culture, it is only a legend or perhaps a reference to some local flood that happened in ancient times. Some even claim that the Babylonian flood story is older than the biblical Flood story. But for those who don't want the Flood story to be true, claiming that because so many flood stories exist somehow discredits the biblical account is only an attempt to hide in plain sight. Perhaps all the peoples of those remote civilizations had different flood experiences that, by chance, had all these features in common, on which they based their stories. But the more reasonable alternative is that these legends all find their root in the same global flood experiences in the biblical account. That so many accounts of a catastrophic flood exist from other sources only gives credibility to the biblical account. As for which flood account is oldest, that seems impossible to prove.

But what we do know is that the biblical Flood account is not a story in isolation. For believers, it's part of a cohesive, progressive story. The innocent animal killed to provide coverings for Adam and Eve is a picture of Jesus, the sacrificial lamb. The ark, a continuation of the human race allowing for a future Savior of the world to be born points to Jesus, the true ark. Only in him are we saved from judgment and death.

Abraham, born six or seven generations from Noah, offered his son Isaac, pointing down through the ages to the time God would offer his own Son, Jesus. All these stories fit seamlessly together and show that not only is the Flood story true, but that it finds its true home in the biblical account.

Many people say that the Jesus of the Bible never existed but that he was just made up, a composite of many people's ideal man. Others say that he did not exist at all, but that many of the sayings credited to him were made up by others and that the miracles were added in later to make him out to be a godlike figure.

The serial killer Jack the Ripper prowled the streets of London for twelve weeks in 1888, savagely killing and dismembering women with a butcher's knife, but he was never caught or even identified. To this day, the police reports detailing his crimes are stored away in Scotland Yard in London. There are names of people whom police suspected might have been the infamous Jack the Ripper. There are photos of four of his victims and autopsy reports on all five victims. There are records detailing the tactics and strategies police used in trying to apprehend him. He was never caught and never brought to trial for his crimes. And his true identity is not known.

In contrast, we know so many personal details about Jesus. We know that his father Joseph was a carpenter. His mother's name was Mary. We know that Jesus was a carpenter himself and that he worked at carpentry until about the age of thirty. We know that he had five brothers and at least three sisters. We know that Jesus was born in Bethlehem, the home of his ancestor, King David. We know that he was brought up in Nazareth.

At the age of about thirty, he began his ministry, telling about and displaying the love of God through his words, his good deeds, and his miracles. We learned that he was no ordinary man but claimed to be the Son of God. We know that the religious leaders of the day opposed him and eventually persuaded Pilate to have him killed by crucifixion. He had twelve disciples, one of whom betrayed him, and we know the names of all twelve disciples. Before being crucified, Jesus was treated shamefully and savagely by the Jews and the Romans. He was nailed to a cross. While hanging on the cross, he said, "Father, forgive them for they don't know what they do." Right up to the end of his life, he was gracious, submissive, and forgiving, in contrast to the dishonesty and mercilessness of his tormentors. We have eyewitness accounts of how he was taken down from the cross and placed in a tomb. Three days later, the tomb was found to be empty. The resurrected Jesus appeared to his followers for the next forty days, getting them used to the reality of his being alive and preparing them to take the resurrection message to the world.

We know all this because of eyewitness reports recorded in Scripture. Jesus's good friend John wrote, "That which was from the beginning, which we have heard, which we have seen with our eyes, which we have looked at and our hands have touched—this we proclaim concerning the Word of Life. The life appeared, we have seen it and testify to it."

The first biography of Alexander the Great was not written until three hundred years after his death. Yet we never hear of any historians lining up to cast doubt on his legacy or claim that he might not even have existed. And yet, people are lining up around the block to try and discredit Jesus's life and work and to cast doubts on whether he even existed at all. Why is that?

At this point, a doubting Thomas might argue, "The Bible has been mistranslated over the centuries." But in universities and museums around the world, there are over five thousand ancient manuscripts, including the Codex New Testament dating to around AD 200. When modern Bible translations are compared to the Codex Bible and to other ancient translations, there is really no difference at all apart from some grammar differences. The modern Bible is remarkably accurate when compared to the oldest known Bibles and transcripts.

Skeptic Thanks

The skeptic would ask, "How could a good God allow so much injustice and suffering in the world?" It's a reasonable question. We are moral beings with the capacity to know right from wrong. The Scriptures provide the answer—God made man good. He gave man the capacity to know and love him. He also gave us maybe the second greatest gift—a free will. God set boundaries for our first parents in the Garden of Eden. He also warned of dire consequences if they went outside the boundaries that he set for them. He told them that they would die. Rebellion entered their hearts; they disobeyed. They hid from God. Why do people hide? Fear would be one reason. Another reason would be guilt. As a result of their rebellion, death entered the world. We, being descendants of Adam and Eve, share the same rebellious nature. "That's not true," you say. Let's do an experiment. What happens if you throw a stone into a pond? The ripples go out in a perfectly formed circle, right? That's a demonstrable fact. In the same way, that we share the same rebellious nature as our first parents is also a demonstrable fact. When Adam and Eve disobeyed God, rebellion and corruption entered their hearts. And that same rebellion and corruption is in all of us, their descendants. Romans 3:23 says, "All have sinned and fall short of the glory of God." Example, we know it's wrong to take harmful substances like drugs because they are bad for our health. Yet so many people still take them. We know it's wrong to cheat, lie, steal, and kill. Yet look at the moral state of our world. It's becoming a more and more miserable place to live in. The vast majority of our problems that cause so much misery and unhappiness are caused, not by religion, not by racial tensions, not by inequality, but by man's sinful, rebellious heart inherited from our first parents.

To say it's because morally we have not evolved fast enough is a complete cop-out. If you follow that logic, any time someone is brought to court for a crime, their defense could be, "I'm running a little behind in my moral evolution; therefore, I need to be pardoned."

Many have a problem with the God morality of the Old Testament. This is a difficult subject to deal with. Smarter and wiser men than I have given this subject a wide berth. Maybe I should do the same, but since I'm not that smart, I will plow ahead anyway.

First of all, I think context—understanding the times in which people lived—is very important. Today, some people think that the laws that Moses gave the people were archaic, excessive, and even immoral. But back then, they were far ahead of any laws of the nations around them.

The Old Testament law stated that if you stole a sheep and were caught, your punishment was to repay four sheep. That may seem tough, but it was better than losing your life. If you took out someone's eye, your eye would be taken out. If you killed someone, you would forfeit your life. I'm not in favor of going back to that kind of justice, but that's where civilization was at that time in their understanding of justice. For all those people out there who are morally outraged, if you kill someone today, you may get seven years imprisonment but get out in four for good behavior. Is that justice for the deceased and their families? It doesn't seem like justice. In order to have justice, the sentence must fit the crime.

In instituting the death penalty, God was putting a high value on life. Equally, God has put a high value on the institute of marriage. That's why the Old Testament laws required those caught in the act of adultery be stoned to death.

In Jesus's time, some Pharisees brought to him a woman caught in the act of adultery. They pressed Jesus to declare if he agreed that she should be punished according to the Old Testament law. Jesus answered them and said, "Let him who is without sin cast the first stone." They all walked away. Was Jesus saying that adultery was okay, that it was not an offense to be judged? The answer is no to both questions because the New Testament tells us that the sexually immoral will be judged. The Bible is very clear on that. But we are living in an age of grace. Because Jesus has paid for our sins, God has deferred judgment, giving us time to turn from our wrong. He is patient, not wanting anyone to perish. But if we don't turn and receive his mercy now in this life, then there will be no mercy on Judgment Day. Don't be fooled. We can't mock God and get away with it. Jesus told the woman, "Go, and sin no more."

Many anti-religious people use the violence of the Old Testament as an excuse to reject God and the Bible. But many stories in the Old Testament are told in narrative form. They are stories reflecting the moral standards of the time. Pick up a newspaper or turn on the TV, and you will see you could well be reading the Old Testament. Man has not changed at all; we are as morally depraved as ever. We have not progressed at all. Our outrage needs to be directed elsewhere—at ourselves, for example. We are told by famous intellectual atheists, "We don't need the Bible to tell us that murder is wrong." Really? Read the news. An awful lot of people seem to think it's acceptable to kill another human being for no good reason other than greed, envy, or personal gain.

When Jesus was nailed to the cross, he was flanked by two others who were also nailed to crosses. At first, both of them mocked Jesus, challenging him to save himself and them if he was God. But later on, for whatever reason, one of them had a change of heart. Maybe he saw the hypocrisy of the religious leaders sneering and cruelly mocking Jesus. Maybe he heard Jesus pray, "Father, forgive them for they don't know what they do." For whatever reason, he shouted at the thief on the other cross, "Leave Him alone. We are getting what we deserve, but He has done nothing wrong." Turning to Jesus, he said, "Remember me when you come into your kingdom."

"We are getting what we deserve." That's an incredible admission. Maybe for the first time in his life, he took responsibility for his actions with no excuses, such as, "I had many disadvantages in my life; I'm a victim of social inequality; I was dropped on my head as a baby." He had no excuses. He was just a man taking responsibility for a wasted life; and he accepted that his circumstances, as horrible as they were, were deserved. What was Jesus's response? He said, "I tell you, this day you shall be with Me in Paradise."

Yes, it's an unjust world. But to whom we attribute the cause tells a lot about what's in our hearts. Corrie Ten Boom was a young Dutch woman who lived in Holland when the Germans invaded her country. She and her family hid Jews in their home to protect them from the Germans who were sending them off to concentration camps. They were found out. As punishment, they were all sent off to concentration camps themselves—Corrie, her sister, brother, and father. Only Corrie and her brother survived. She suffered unspeakable horror at the hands of the Germans, but she survived. Later in life, she forgave the Germans for what they did to her and her family and spent the rest of her life traveling the world, telling about God's love and grace until she died at the age of ninety-one. She didn't blame God for all the bad things that happened to her in her life.

One time, while corresponding with my atheist friend, I asked him, "What if God really does exist? What would you say to him if you met him?" His response was, "I would give him a piece of my mind." I wrote back to him and asked, "Really? Have you nothing to be thankful for? Don't you have a lovely home in a good location? Don't you have relatively good health, three meals a day, and your dog for company?" After a delayed reply, he wrote back, saying, "I hate my life. I wish I were never born." Before anyone sheds a tear for him, he enjoys life and makes the most of it. Only a few weeks earlier, he had been talking about how he loved his time working in London. It's just that he would rather distort the truth than acknowledge that he had something to be thankful to God for. He wants to go on hating God. Like many others, that's the only narrative he is interested in.

The famous Naturalist David Attenborough was once asked if he believed in a creator. He paused for effect, and then went on to talk about a fly that feeds on the blood of people and in the process lays it's eggs in the same host. Parasitic worms, Onchocerca volvulus, hatch from the eggs then proceed to live and breed inside there human host, causing terrible itching and the unfortunate person to break out in spots all over. Eventually some of the larve make there way to the persons head and eyes where they damage the cornea causing the person to eventually go blind. Mr. Attenborough went on to say that he could not believe in a god who would create such a creature that caused children to go blind.

The fly he was talking about is known as 'The Black Fly.' It's found in about 30 countries in Africa and in some countries in South America. It's favored habitat is near rivers where it lays it's eggs. Unlike the mosquito which has a probosus that it uses to penetrate through the skin, the black fly has two slicing blades that it uses to cut a hole in the skin. Then, while feeding, larve fall from the black flies mouth onto the pool of blood where they make there way into the host, breed and multiply.

A terrible disease for all those who have to endure it. And not only that, but prime land where these flies live and breed uncheacked, has to be abandoned, or else people face the very real danger of becoming infected.

But what David Attenborough failed to point out is, 'that's not the end of the story.' In 1994 the WHO began a control programe where they sprayed insecticide into rivers where the black fly lives and breeds. The programe was so successful that it allowed many people to return to their homes and farmland. And the story doesn't end there. Later on it was discovered that invermetin, a drug used to treat heartworms in dogs, worked so effectively against river blindness that it was almost a miracles drug, killing the larve in the human body and stopping the itching almost immediately. And most importantly, putting an end to people going blind because of the black fly. And done with just one or two pills per year per person. The success is so unprecedented there is now every chance river blindness will be eliminated completely.

David Attenborough, a man who by any measurement seems to have lived a very rewarding and fullfilling life, observed a tragedy and made a moral judgement on it. His judgment, I can't accept a god who would allow such tragedy. The tragedy didn't seem to have rendered him unable to function or live a normal

CREATION OR EVOLUTION: A LAYMAN'S LOOK | 51

life. When Mr. Attenborough views these things from the lense of evolution, his simply accepts the evolutionists explanation for suffering and death: 'it's the survival of the fittest. Natures way of weeding out the weak.' But then, when viewed from a God and morality persective, he puts on his morality cap and stands in judgement of God. I could understand his reasoning completely if I, like him, accepted the evolution theory and rejected the revelation of scripture. Evolutionists claim, 'that's just the way things are.' Scripture says, 'no' it wasn't always that way. Man rebelled and tragedy and death came into the world. But that's not the end of the story. One time when thinking on some of the gospel stories I had read, the thought came to me, "if Jesus could feed fifteen thousand people with five loaves and two fishes, then he could feed the whole world. If he could heal some crippled or diseased people, then he could heal all the sick or diseased. And some day he will. He will do away with all sickness and death.

Fanny Crosby was a famous hymm writer. She was born in the US in 1820. She wrote more than eight thousand hymns and gospel songs. She only began writing songs in her 40's and for the rest of her life wrote about three songs a week. She lost her sight when she was six years old. The result of a doctors error in treating an illness she was suffering from. She never expressed any bitterness. In many of her songs she writes about seeing her Savior someday. She viewed her blindness as a tempory disability.

I have nothing against David Attenborough. I think he is brilliant and I love his programes. But he, like many, needs to open his eyes to the whole story. We can only do that when we reject the lie and accept the truth. Sir Isaac Newton wrote the song Amazing Grace. In it he says, "I once was lost but now am found. Was blind but now I see."

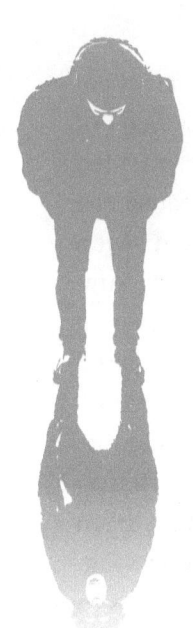

Abandon Ship

It seems like never before, the world is drowning in a tsunami of cynicism. Ancient-held beliefs that have served us well for millennia are under attack. These attacks are not selective. All of our dearly held beliefs and values, in the minds of those who want to leave us rudderless, sailing into the abyss, need to be torn down and done away with. They claim we need to start over again in order to have a just and fairer world where everyone is treated equally. But they themselves are the most intolerant and manipulative people, so how fair and equitable a world could it be with them in charge? And almost without exception, the people behind these movements want God out of our institutions and public life. They tell us that God doesn't exist, that we are alone in the universe and masters of our own destiny. They dedicate their lives to destroying trust in God. The irony of it is that in doing so, they spend most of their time thinking about God. And yet, they want you and me to stop thinking about Him.

People put forward different systems of government that they say will lead to more egalitarian societies; but no matter what system of government is implemented by the people, they don't work, mostly because the system isn't the problem. Man is the problem. Man makes perfectly good laws, and then when it suits him, he puts himself above the law or tries to find ways to circumvent them. We could flatter ourselves by saying we have laws because men and women are sophisticated beings who live in complex societies of our own making. But the other equally pertinent, but not-so-flattering reason is, man is a lawbreaker. We know right from wrong; but because we are so corrupt and self-serving, if we have to, we are willing to operate outside the law in order to have an advantage over others and to get ahead.

Someone has said, the main reason we have about eight thousand laws is because we can't even keep the ten laws God gave us on Mount Sinai. God's laws are spiritual, but we are not spiritual. That's why we can't keep them.

One of the Beatitudes is "Blessed are they who know their spiritual poverty." There is the real reason we don't come to God. It's the reason we reject religion. We think we are fine as we are, that we just need to work a little harder to improve ourselves, not realizing we are poor, miserable, and naked. When others do wrong, we judge them and often despise and avoid them. However, when we commit the same or similar offenses, we make light of our inconsistencies, make excuses, and give ourselves the benefit of the doubt. We are easy on ourselves and hard on others when it should be the other way around.

Run, Rabbit, Run!

One of my favorite movies is *Cool Hand Luke* with Paul Newman. In it, the prison warden, who was as corrupt as Luke or any other of the prisoners in his charge, was lecturing the prisoners about the consequences of trying to escape. He used an illustration I always thought was a very insightful way to describe human nature—"Man's got rabbit in him." Luke and the other prisoners were there because while they were growing up, they had made wrong choices. They failed to deal with their rabbit. Like the prison warden who failed to see that he himself had rabbit in him, we need to recognize and deal with the rabbit in us before it's too late.

The world is the way it is but not because of God. We fail to recognize just how warped and destructive our natures are and that only by turning to God can we be saved from ourselves.

We think science is the answer to solve our own problems. But time is running out. The Russian playwright Anton Chekhov (1860–1904) said, "Man has been endowed with reason, with the power to create, to that he can add to what he has been given. But up to now he hasn't been a creator, only a destroyer. Forests keep disappearing; rivers keep drying up; wild life becomes extinct; the climates are ruined; and the land grows poorer and uglier every day."

We need to do what few men have ever done—stop running away and really look into our hearts. Jeremiah 17:9 says, "The heart is desperately wicked and deceptive above all else. Who can understand it?" It took men ten years to build the World Trade Center. It took only one hour to knock them down. Yet man stubbornly insists that he and he alone can solve the problems of mankind.

The American writer F. Scott Fitzgerald said, "I refuse to accept the end of man. He will prevail because, alone among all creatures he has a soul, a spirit, capable of compassion, endurance and sacrifice."

These sentiments I agree with, but what should we do about that rabbit we so unwisely overlook?

A chain is only as strong as its weakest link. We are only as strong as our weakest moral point. Watching the news is an absolute horror show. Man is a predator without equal in all of creation. For the last ten years or so, we have witnessed in real time millions of people forced from their homes, their countries. Man-made famines are on a scale rarely seen in history. Corruption is so vast and so ingrained that it's virtually impossible to root out. Conflicts arise between nations, communities, and races. Most of these problems can be traced to the corruption in man's heart. Mankind is being pushed to the brink. All the vileness in the heart of man is pushing us to the edge of a cliff and the momentum being generated by all of these crises will push us over. Have you heard of the Doomsday Clock? It was started in 1945, when people like Robert Oppenheimer and Albert Einstein formed the Bulletin of the Atomic Scientists. The clock itself was designed in 1947 by an artist named Mary Langsdorf. The clock is a symbol for the threat of an impending nuclear apocalypse. Scientists take into account other nonnuclear factors, like climate change, bioweapons, and cyber threats. In 1953, when the US developed the hydrogen bomb, the clock was set at two minutes to midnight. But it was put back to twelve minutes to midnight when the US and Russia ended atmospheric-nuclear testing. Currently, it is set at one hundred seconds to midnight.

How bad do things have to get before we realize that not only does man not have the answers to the problems but that we are the problem.

One sin brought death to all men. And because of that, one righteous man had to die to pay the price for our sin. We need to stop making excuses and stop trivializing sin. No, we need to make a beeline to the cross and be saved from sin.

A Real Man

Most of us have heard the Bible story about Joseph. He was the favorite son of his father Jacob, which caused his brothers to hate him. When the opportunity came, they decided they would kill him. But they changed their minds and instead sold him into slavery. In order to trick their father into thinking that Joseph had been killed by a wild animal, they took Joseph's multicolored coat and dipped it in blood. This caused their father to give up on ever seeing Joseph again.

After enduring a difficult time for many years, Joseph was able to interpret the Egyptian king's dreams. As a result, he was promoted to second ruler of all Egypt. With revelation from God, Joseph had predicted a seven-year famine, which, sure enough, came on all the countries in that region at that time.

Since the Egyptians knew that the famine was coming, they were able to prepare for it; but the rest of the world was unprepared, including those in Canaan, Joseph's homeland. Jacob sent ten of his sons to Egypt to get grain to feed his family. After many years in Egypt, Joseph was no longer a youth. Now he looked and spoke like an Egyptian, so his brothers didn't recognize him. When Joseph saw his brothers, he spoke harshly to them and accused them of being spies. But Joseph's brothers said no, insisting that they were not spies but claiming that they were good and honest men. Imagine that—they stood before their brother, whom they had sold into slavery and claimed that they were good men. They also lied to their father and caused him great grief, and yet they claimed that they were honest men.

To make a long story short, Joseph revealed himself to his brothers. For a long time, they were so stunned that they couldn't speak. When they finally found their voices, they all talked and cried for a long time. Joseph forgave them, telling them that God had sent him ahead to save them. In the Old Testament, Joseph is a type of savior. He suffered for a while and then was in a position to save his people, pointing to the time God would send His Son, Jesus, to be the Savior. Of course, the parallels between Joseph and Jesus differ in that Jesus really is the Savior of the world, and he really did die and come back to life from the dead.

Like Joseph's brothers, who were speechless before him and didn't deserve his love and forgiveness, we are all guilty before God and don't deserve his love and forgiveness. Deep down you know this to be true. Someday when we stand before God, we will all be like Joseph's brothers, speechless and guilty. Unless we turn to the Savior God has sent to us. God sent Jesus to die on a cross for us so that we can be forgiven. God loves us. He proved it by sending Jesus to die in our place. Like Joseph's brothers, the only thing we can do is receive his love and forgiveness. It would be foolish not to. Joseph's brothers had the good sense to receive his love and forgiveness.

Proud and haughty man! You who beat your chest and shake your fist at God! Listen to how God answered Job when Job accused him of wrongdoing. God said to him, "Can you run with horses? Can you put your hand into a crocodile's mouth and not regret it?" What God means is there are flesh and blood creatures that are too powerful and too fierce for us to contend with. So how can we contend with Almighty God? Jesus said, "Make peace with your enemy while on the way to court with him. Otherwise he may hand you over to the judge. And the judge may throw you into prison. I tell you, 'you will not get out until you have paid the last penny.'"

Don't be foolish. Come to your senses and make peace with God.

> In the words of an old hymn,
>
> Come to the Lord where life in its fullness is found. And let Him save you the way He's already saved me. You'll find peace for your soul, and He'll make you His very own friend.
> And He'll give you a mansion in heav'n when this life is through.

SMALL MINDS

One time when discussing with a group of people, the Apostle Paul asked the question, "Why do you think it impossible that God could raise the dead?"

We are conditioned by the laws of nature to know that if we throw something up in the air, it will come back down again. If we drop something heavy on our foot it will hurt. In the same way, nature has taught us that when someone dies, that's it. Death is final.

We defer to logic and reason to help us understand and make sense of everything around us, including creation itself. We prefer to think that somewhere in the very distant past, billions of years ago, the beginning of our material world was birthed. Of course, how that happened, no one can explain. And that, somehow since then, through trial and error, creation has miraculously gotten to where it is now through an almost infinite number of trials and errors.

Evolutionists tell us these things. And they tell us in such a way that it's all neatly put together like a meal in a fast-food restaurant. We just accept it without really thinking for ourselves. When some break ranks and do ask questions, they are told, "It's too complicated for you."

And not only are evolutionists sure of how creation came into being, they are equally or even more sure that a supernatural god had no part in creation. Or that such a being even exists.

Once while observing some ants at work, I began to think about how limited their understanding is. They went about in their own little world, oblivious to the wider world around them. They couldn't comprehend me, for example, or the motor vehicles going back and forth on the road nearby. Humans are far superior to ants.

But although we are far superior to the ant, like the ant, we too have limited understanding. Like stated above, experience has programed us to stay within the limits of nature's laws. When we think about God, we think of him as being like us, only more superior. But God, as revealed in the Bible, is not flesh and blood as we are. In Psalm 139, David says, "Lord, where can I go from your Spirit? If I go to heaven you are there. If I go to the deepest depths you are there." God is Spirit, not limited to one place like us.

In the Old Testament, God told Moses, "You cannot see My face and live." The reason given is that God is holy. And that is true. But there is another just as equally mind-blowing reason. God is so superior to us that we simply don't have the capacity to comprehend him.

God is not like anything in his creation, including us. In that sense, God is unknowable. If we want absolute empirical proof that God exists, then we want something that is beyond our capacity as finite beings. Just like an ant could never comprehend us, so we could never fully understand and know God. Even if a seemingly superior being was to appear and say, "I am God," how could we prove it? We are finite and limited in our capacity. We could never prove the eternal.

Satan wanted to be like God and that caused his downfall. He tempted Eve with the promise that she would be like God and so caused her downfall. Insisting that we must understand God before we believe in him is just arrogance. Believing that we can understand everything is foolish.

So we need to dial into God's way of making himself known to us. Of course, nature is a powerful proof of a creator. But the life and words of Jesus are a still greater proof. Jesus claimed that he "was sent by the Father and that his kingdom was not of this world." Before he died, he claimed that he had the power to lay his life down and the power to take it back up again. After he rose from the dead, he was seen by over five hundred people. That's why Paul asked the question, "Why do you think it impossible that God could raise the dead?"

If we can convince ourselves that our world and universes beyond our world came into existence billions of years ago, then why do we think it impossible that God could make all things in six days?

When God spoke audibly to Jesus in the hearing of others, some of those who heard God's voice said,

"It thundered." It is easier for us to relate to natural phenomena than to recognize and accept supernatural phenomena.

Jesus foretold that Jerusalem would be destroyed. Two days before he was crucified, as he rode into Jerusalem on a donkey, he stopped and wept over the city, saying, "Jerusalem, Jerusalem, if only you had known what would bring you peace. But now your house is left to you desolate." Forty years later, the Romans attacked and destroyed the city, killing over 1 million people. Historians would tell us all about the political and social conditions leading up to the attack by the Romans. They would tell how the Jews were not happy with Roman rule, how there was unrest and rebellion because of high taxes, and how the Romans lost patience with the Jews leading them to attack and destroy Jerusalem. But nowhere in the history books does it tell about how Jesus foretold what would happen or that things could have been so different if the Jews had embraced their Messiah.

In conclusion, we have the creation story or the evolution theory. Both are fantastical and impossible to understand, but as far as reason and thought can be applied here, the creation story is the more believable.

Jesus said, "Heaven and earth will pass away. But My Words will never pass away."

Personal Testimony

Certain prominent atheists have likened heaven to a celestial North Korea. But that is a complete misrepresentation of heaven as described in the Bible. The Bible describes heaven as a place where there will be no more tears, no more illness, and no more death. There will be no more conflict because residents of heaven will have a new nature. Instead, our wills will be completely in line with the Father's and submitted to his will. Free from the conflict caused by our rebellious nature, we will, instead, live in harmony with our surroundings and with each other with Christ as the center. In the gospels, we have a picture of what a life free from the conflict of a rebellious nature is like.

When we look at Jesus, everything else is out of focus because he is incorruptible in thought, word, and deed. Everything was his, but he laid it aside. Born in a stable, born in humble circumstances, tested to the limits by a hostile world, tortured and hanged on a cross, yet he responded by praying, "Father, forgive them because they don't know what they do." It's the world that is out of focus. We are out of focus. Not Jesus. Not God's Word.

Everything is his, but he allows us freedom of choice. Unlike earthly dictators who force people to submit to them. Instead, Jesus allows us to make our own choices, to accept him or to reject him.

He said, "Come to Me all you who are weary and burdened and I will give you rest. Take My yoke upon you and learn from Me, for I am gentle and humble in heart; and you will find rest for your souls. For My yoke is easy and My load is light." Those are not the words of a dictator. Instead, Jesus showed time and time again that it wasn't earthly power he was after. Or control over others. Satan offered him the kingdoms of the world and all their glory. He turned them down. The people wanted to make him king. He refused. In the book of Revelations, Jesus is recorded as saying, "Behold, I stand at the door and knock. If anyone answers I will come in a sup with him and he with Me." Jesus waits for us to open our hearts to him and answer his call. Dictators don't knock and politely wait for you to answer. They just kick the door down.

In the thousands, multiple thousands, people came to Jesus for help. For hope. And he didn't disappoint. But by the end of his earthly ministry, he had only about five hundred committed followers. He said himself, "Broad is the road that leads to destruction and many there are who take it. But narrow is the road that leads to life and only a few find it."

Those on the broad road live for this life only and are self-serving. They want to maintain lordship over their own lives and answer to no one. They don't reject God completely, but neither do they submit to God's claims on their lives. Some do reject God completely and are even openly hostile toward him, like the atheists who claim heaven is just a celestial North Korea and God the evil dictator. Maybe you think, "It's all just one big mistake on their part. And if they knew where they were wrong, then they would be quick to correct their mistake." No! Not at all! They know full well they are misrepresenting the gospel, smearing God's good character, and rejecting Jesus's claim on their lives.

They also claim they don't like the idea of a God who polices their every thought. The gospels all tell about how Jesus predicted he would be arrested and that his followers would desert him. It happened just as he predicted. Peter, however, followed at a distance right up to the courtyard where Jesus was detained. Three times he was challenged about being one of Jesus's followers. Three times he denied even knowing Jesus. One of the gospels records how, at the moment of the third denial, Jesus turned and looked straight at Peter.

Yes, God knows our every thought before we think them; he knows our every action before we act them out. But no matter what we think or do, God still loves us even in spite of ourselves. Like the song says, "What a friend we have in Jesus, all ours sins and griefs to bear. What a privilege to carry, everything to God in prayer." God is not waiting with a stick to beat us when we think or do something wrong.

Many years later, when the Romans sentenced Peter to death by crucifixion, Peter asked that he be crucified upside down because he did not consider himself worthy to be crucified as his Lord was. That's what Jesus's love did for Peter. Peter didn't feel stifled by a God, who knew his every thought and motive. Instead, he felt grateful to a God who loved him in spite of himself.

In the book of Jeremiah, it says, "For I know the plans I 'have for you,' declares the Lord, 'plans for good and not for evil, to give you a future and a hope." That's the heart of God toward us. He is not at all like some would have us believe, waiting to pounce when we do something wrong. Instead, when we come to him, he opens our eyes to our true character and, more importantly, his true character. And then he shows us the solution: Jesus, the One who never thought, said, or did anything displeasing to God. The One who died to pay for our sins and rose again so we could have eternal life. Some accept this priceless gift of God, but most reject this offer of hope in Christ.

Before becoming a believer, I was religious and thought about God from time to time. If I ever did a good deed—and they were rare—I would be pleased with myself and offer it to God as evidence that I was a good person and not all bad. But mostly I was a very selfish, self-centered person and lived to satisfy my own needs and wants. But personal issues, caused in large part by my rebellious and depraved lifestyle, caught up with me. Unable to cope with life and not knowing where to turn for help, I began reading a New Testament one of my sisters had gotten in school. It seemed almost every time I opened it, it spoke about peace. And peace was something I didn't have at the time. I felt God's presence in my life like never before. My outlook changed overnight. I felt a hope and a purpose that I'd never felt before. God's presence in my life was very real and very personal. One day while thinking about what God's will for me was, I had a revelation from God. He brought to my mind the story of Noah's ark, which I had recently read about in the Bible. In it, Noah and his family went into the ark, the rains came, and everyone outside the ark perished. God showed me very clearly that Jesus is the true ark and that anyone in him is saved from God's judgment on sin. I went up to my room, got on my knees, and prayed, "Lord Jesus, I thank you for dying for my sins. I accept you as my Lord and Savior." I believe I was accepted by God before I actually prayed that prayer because I had a peace and a hope in my life that I never knew before. So it was easy to pray that prayer and mean it.

As I continued to read the Bible and walk with the Lord, he revealed to me more and more my true condition. He was showing me just how corrupt my heart really is. It was a very uncomfortable and embarrassing experience. When you open yourself up to God, shining his light into your soul, there is nowhere to hide. And neither did I want to hide. I wanted everything out in the open and dealt with. I was no longer comfortable living a lie or making excuses for my behavior. And I never felt condemned by God. Embarrassed before him, yes. But never condemned. Why not? Well, because all my sins have been forgiven and paid for by Jesus on

Calvary's cross. No condemnation. Just God showing me the reason for all the unhappiness in my past life and freeing me to live this new life.

I'm glad God knows everything about me. How else could he help us if he didn't? He is not waiting to condemn. He is waiting to forgive and make new. In the story of the prodigal son, Jesus gives us a picture of what God is really like. The prodigal son took his inheritance, traveled to a distant land, and wasted all he had in wild living. Then a famine came on the land, and he didn't even have enough to eat. His need caused him to have a change of attitude, and he came to his senses. He decided to return home to his father and ask his forgiveness. The story tells us that, while he was still a long way off, the father ran and embraced him, welcoming him back. We are all sinners, therefore, we are all prodigals.

Before coming to faith, I lived life according to my own standards. The problem with that was, I had very low standards. And even they were negotiable if it suited me or helped me get what I wanted. Now, when I sometimes look back on my life at the shameful things I have done or said, I cringe. But that's not God condemning me. I know that God accepts me and that all my sins have been forgiven because of Jesus's sacrifice. Jesus didn't come to condemn. Instead, he said, "I have come that you may have life and have it in its fullness."

www.ingramcontent.com/pod-product-compliance
Lightning Source LLC
Chambersburg PA
CBHW021015180526
45163CB00005B/1955